DK ESSENTIAL SCIENCE

killer diseases

HAZEL RICHARDSON

SERIES EDITOR JOHN GRIBBIN

LONDON, NEW YORK, MUNICH,
MELBOURNE, and DELHI

series editors Peter Frances and Hazel Richardson
DTP designer Rajen Shah
US editor Margaret Parrish
category publisher Jonathan Metcalf
managing art editor Philip Ormerod

Produced for Dorling Kindersley by
Grant Laing Partnership
48 Brockwell Park Gardens, London SE24 9BJ

managing editor Jane Laing
managing art editor Christine Lacey
picture researchers Jo Walton, Louise Thomas
illustrator Peter Bull
indexer Dorothy Frame

First American Edition, 2002

02 03 04 05 10 9 8 7 6 5 4 3 2 1

Published in the United States by
DK Publishing, Inc.
375 Hudson Street
New York, NY 10014

Library of Congress Cataloging-in-Publication Data

Richardson, Hazel.
  Killer diseases / Hazel Richardson.
    p. cm. -- (Essential science)
  Includes bibliographical references and index.
  ISBN 0-7894-8922-8 (alk. paper)
    1. Epidemics--Popular works. 2. Epidemiology--
  Popular works. I. Title. II. Series.

RA653 .R53 2002
614.4--dc21
                                    2002071499

Color reproduction by Colourscan, Singapore
Printed and bound by Graphicom, Italy

See our complete product line at www.dk.com

P9-DVT-966

contents

our greatest threat

Millions of people have been killed throughout the ages in wars, accidents, and natural disasters. Horrific though this is, the numbers lost are nowhere near the casualties due to killer diseases. In the last 100 years or so, we have discovered effective treatments for numerous illnesses, and many of the most dreaded have become less of a threat to us. However, diseases still kill several million people around the world every year, and new ones are constantly appearing. The most intense threat to our future comes from the unseen microorganisms that surround us. Bacteria, viruses, protozoa, and other parasites are constantly assaulting our bodies, held off only by our immune system. When our defenses fail, we become ill, sometimes incurably. Lately, especially in the Western world, our disease burden has been increased by our own actions. Unhealthy diets, drug abuse, lack of exercise, tobacco smoking, and greater sexual activity have led to a drastic rise in cases of killer diseases, including heart disease, lung cancer, and AIDS.

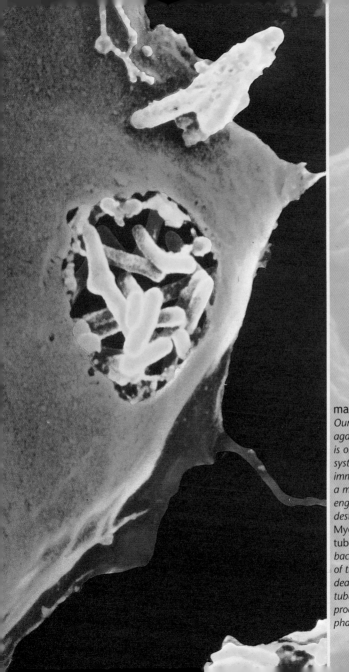

macrophage
Our main defense against disease is our immune system. Here, an immune cell called a macrophage is engulfing and destroying Mycobacterium tuberculosis *bacteria—the cause of the potentially deadly disease tuberculosis—in a process known as phagocytosis.*

the causes
of disease

If asked what disease is, most people would answer that it is anything that makes somebody sick. Although this is a good answer, it is not the full story. Many diseases can take months or even years to result in symptoms, even though they are causing irreparable damage to the body during this time. Some diseases have mild effects, while others can kill. And diseases affect all living things on Earth, even bacteria.

The standard definition of disease is any condition in which the healthy functioning of a living organism is disrupted and in which there is a physical change in its cells or tissues. Whether the disease is mild or serious depends on the level of disruption. For example, a certain amount of the chemical vitamin C is needed by the human body to maintain healthy bones and blood vessels. A mild vitamin C deficiency may cause no symptoms, or result in aching and nosebleeds; a larger one may lead to anemia and scurvy.

The seriousness of disease is also related to the extent and the site of any physical damage. An infection of the skin can cause scarring but is usually not life-threatening, while damage to heart muscle may result in a fatal heart attack or heart failure.

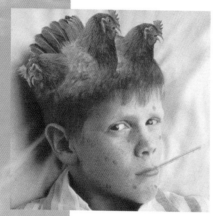

chickenpox
A common childhood disease, chickenpox is normally a short-lived illness, causing a rash and slight fever. It is rare in adults, but when it does occur it tends to have more severe effects, including pneumonia.

types of disease

There are two main classes of disease: exogenous and endogenous. Most human diseases are exogenous—they have an external cause, such as exposure to a toxic substance, injury, radiation poisoning, or infection. Endogenous diseases are those that arise within the body and do not have an external cause. Parkinson's disease, caused by the death of nerve cells in the brain, is one example. Endogenous illnesses cannot be directly transmitted to other people—they are not infectious. However, some of them, such as genetic disorders, can be passed on to an affected person's offspring.

Whether a disease is exogenous or endogenous is not always clear. The causes of some illnesses, such as multiple sclerosis, are still unknown. Many diseases, such as some cancers, that were thought to be endogenous in the past, are now known to have an external cause or trigger, such as a previous infection with a virus or exposure to a toxic chemical.

danger from outside

Our bodies are constantly assailed by external threats to health. Some of these have minor effects; others result in millions of deaths every year. Causes of exogenous disease include radiation (either naturally occurring or from nuclear weapons or power stations), toxic chemicals including poisons, carcinogenic (cancer-causing) substances such as asbestos and tobacco, and the effects of infectious agents.

Infectious agents are by far the most frequent causes of disease, and the deadliest foes we face. Apart from the recently discovered prions, which cause the rare brain diseases Creutzfeldt-Jakob disease and kuru (see pp.64–65), all infectious agents are parasites. These live in or on a host and use it for food and to reproduce. They range in size

poisonous substances
Any substance that disrupts the functioning of the body when taken in relatively small amounts is poisonous. Poisons can be swallowed, inhaled, absorbed through the skin, or injected under the skin.

from microorganisms that are too small to be seen by the human eye, such as bacteria and viruses, to insects such as lice and fleas, and other creatures, such as worms.

living off a host

Fleas, as well as lice, ticks, mites, parasitic flies such as mosquitoes, and some fungi and bacteria, are ectoparasites. Human ectoparasites, which may spend all or only part of their lives living on us, get their nutrients from the skin or from body fluids such as blood. They rarely cause serious disease by themselves, usually just slight symptoms, such as itching or lethargy. (The term "lousy" comes from the feelings suffered by people infested with lice.) Some, such as the mites that infest our eyelashes or our beds, living off our discarded skin flakes, are not noticed at all. Ectoparasites are usually a major threat to our health only when they carry endoparasites that they pass on to us.

Endoparasites live inside their hosts. They include all viruses, many bacteria, a few fungi, and some species of protozoa and worms. Some can live and reproduce only in host cells; others infect a particular organ or region of the body, such as the intestines or the bloodstream. They have killed more humans throughout history than anything else. Just some of the human killer diseases they cause are malaria, bubonic plague, cholera, sleeping sickness, yellow fever, and tuberculosis.

❝ They [viruses] make themselves known by the cells they destroy, as a small boy announces his presence when a cake disappears. ❞

Max Delbrück, German biophysicist, 1945

the alien within
This schistosome fluke has left the freshwater snail that it had parasitized to burrow under the skin of a person swimming in infested water. Inside its human host, this endoparasite causes the development of the debilitating disease schistosomiasis.

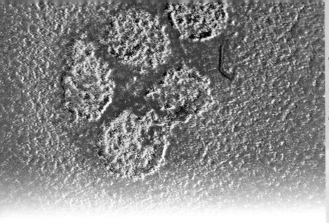

yellow fever
This electron micrograph shows yellow fever virus particles. The virus is transmitted by mosquitoes and may cause jaundice (yellowing of the skin)—hence the name. Sufferers may also develop kidney failure and delirium, which can lead to coma and death.

how parasites cause disease

The symptoms of a disease caused by a parasitic organism are mainly due to the effects of the organism, destroying cells or tissues, releasing toxins, or drawing on the person's supply of nutrients. The effects are slightly different for each organism, which can help when diagnosing infectious diseases.

Viruses are the ultimate parasites. Scientists are still debating whether they are actually living organisms at all, or whether they are just collections of large molecules that can replicate themselves in the right conditions. The sole function of a virus is to invade a cell and take over its DNA to use it for its own purposes. This prevents the cell from working normally. Sometimes, the alteration of the cell's DNA by the virus causes cancer to develop (see pp.54–55). At some point, the virus uses the host cell's DNA to replicate itself, creating thousands of new virus particles, which break out of the cell, killing it. These virus particles then go on to infect thousands more cells, and the cycle begins again. Viruses are responsible for nearly all of the most horrific killer diseases. Herpes, smallpox (thankfully now eradicated), polio, hepatitis, the flu, encephalitis, mumps, measles, Lassa fever, Ebola, rabies, some cancers, and Acquired Immunodeficiency Syndrome (AIDS) (see p.56) are all viral diseases.

infectious organisms

viruses

These are the smallest infectious organisms. They consist of a single or double strand of genetic material encased in a protein shell. They occur in three main forms and can only reproduce inside a living cell.

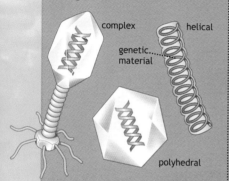

complex · helical · genetic material · polyhedral

bacteria

These are microscopic, single-celled organisms that have a variety of shapes, which are used to help in classification. Some have hairlike projections called flagella, which are used for movement. Most have rigid cell walls.

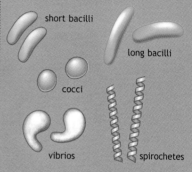

short bacilli · long bacilli · cocci · vibrios · spirochetes

human diseases
- influenza
- yellow fever
- herpes
- AIDS

human diseases
- tuberculosis
- leprosy
- pneumonia
- syphilis
- scarlet fever
- typhus

HIV-infected T cell

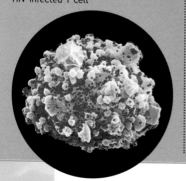

tuberculosis bacteria

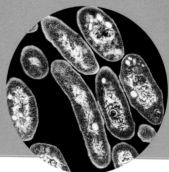

protozoa and worms

Protozoa are single-celled organisms that can parasitize humans at some stage in their life cycle. Worms range from microscopic to several yards long. There are two types: flatworms, including flukes, and roundworms.

fungi

Fungi are single or multicellular organisms. There are two main groups that cause human disease: filamentous, which form branches that invade tissue, and yeasts. Many have dormant forms called spores that can establish infection if inhaled.

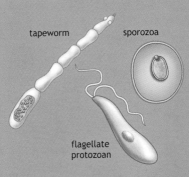

tapeworm

sporozoa

flagellate protozoan

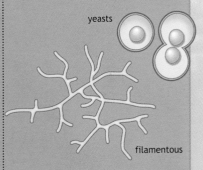

yeasts

filamentous

human diseases
- malaria
- sleeping sickness
- schistosomiasis
- liver-fluke infestation

human diseases
- ringworm
- athlete's foot

liver fluke

trichophyton mentagrophytes (athlete's foot)

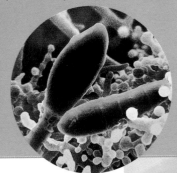

key points

• Diseases can have an external or an internal cause.
• All infectious agents, except prions, are parasites.
• Only infectious diseases can be spread between people, although genetic diseases can be passed on to offspring.

A few bacteria damage tissues by directly invading cells and multiplying within them, but many produce toxins that destroy cells or stop them from working properly. The botulism bacterium produces the most powerful poison known—a quarter of a gram could kill about 60 million people. Intestinal worms cause malnutrition by digesting food we eat. Protozoa cause disease by producing toxins or by invading cells and reproducing in them.

danger from within

Many endogenous diseases occur as an effect of natural aging. No one is yet sure exactly why we age. Some scientists claim that our cells can divide only a certain number of times without error, others that our bodies are slowly poisoned by chemicals we consume and produce during metabolism. Perhaps aging occurs because of a combination of these events.

The major effects of aging are degeneration of nerve cells in the brain, a reduction in the effectiveness of the senses, less efficient working of the heart and lungs, hardening of the arteries, joint problems, loss of bone density, and less efficient liver action. There is also a general weakening of the immune system, which usually protects us against infections. This means that the elderly are more likely to suffer fatal effects from infections such as influenza, which in earlier years they would have survived. The likelihood of developing cancers and heart disease also generally increases with age.

hard of hearing
As we get older, cells in our inner ear responsible for hearing degenerate or are damaged, with the result that at least some degree of deafness is common in the elderly.

how infections occur

We are surrounded by viruses, bacteria, and fungi, some of which are capable of causing human disease. Although just one bacterium or virus entering our body and then reproducing could theoretically cause disease, infections do not occur this easily. This is because we have a defense mechanism that prevents infectious agents from entering our body and quickly kills those that slip through. It is called the immune system.

defense against disease

The human immune system has three levels to protect against infection. The first is barrier immunity, which prevents the microorganisms that surround and cover us from entering our body in the first place. Skin is an effective barrier to most microorganisms. It also contains sebaceous glands. These secrete chemicals that are highly poisonous to many bacteria.

Places in the body where organisms could enter—the eyes, mouth, nose, vagina, and urethra—have additional defenses. Tears produced by lacrimal glands next to the eyes help to wash microorganisms off the eyes; they also contain an enzyme called lysozyme, which can digest bacteria in the same way as enzymes in some laundry detergents digest food stains on clothes. Lysozyme is also present in saliva in the mouth.

under attack
We are in constant danger of contracting an infection. Our skin is covered with microorganisms; our food and beverages are infested with them; and they float unseen in the air all around us, waiting to be breathed in.

how the immune system works

When barrier immunity is breached, white blood cells provide two further lines of defense. The first is the inflammatory immune response (below), which involves phagocytic white blood cells engulfing foreign organisms (phagocytosis). The second is the specific immune response (right), which has two components: the antibody and the cellular responses.

.... infection enters body
through broken skin

phagocyte   blood
vessel wall

The damaged tissue releases chemicals that attract phagocytic white blood cells and make blood vessel walls more porous.

The phagocytes engulf and digest foreign organisms.

More phagocytes, attracted by chemicals, move into the tissues through the porous blood vessel walls.

white blood cells

Individual types of white blood cell have different functions. Neutrophils and monocytes carry out phagocytosis. Lymphocytes provide specific immunity. Basophils release chemicals that cause inflammation, and eosinophils are associated with allergy.

monocyte

basophil

neutrophil

lymphocyte

eosinophil

antibody response

Phagocytes travel to a lymph node and "present" processed antigen from engulfed microorganisms to B lymphocyte cells. These enlarge and divide, developing into plasma cells, which, in turn, release antibodies that seek out and lock on to antigens, inactivating the foreign organisms.

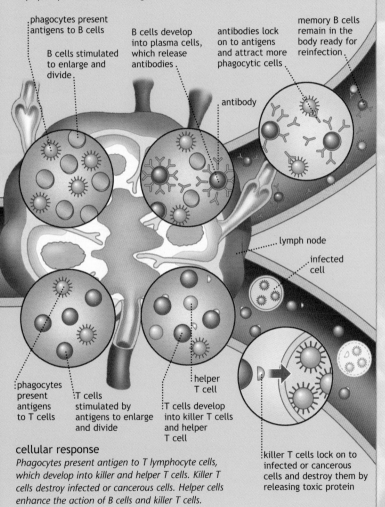

phagocytes present antigens to B cells

B cells stimulated to enlarge and divide

B cells develop into plasma cells, which release antibodies

antibodies lock on to antigens and attract more phagocytic cells

memory B cells remain in the body ready for reinfection

antibody

lymph node

infected cell

phagocytes present antigens to T cells

T cells stimulated by antigens to enlarge and divide

T cells develop into killer T cells and helper T cell

helper T cell

killer T cells lock on to infected or cancerous cells and destroy them by releasing toxic protein

cellular response

Phagocytes present antigen to T lymphocyte cells, which develop into killer and helper T cells. Killer T cells destroy infected or cancerous cells. Helper cells enhance the action of B cells and killer T cells.

Parasites that do manage to get into the digestive system are usually destroyed by acid in the stomach. If any bacteria do make it through to the intestines intact, they find billions of harmless bacteria (called commensals) lying in wait for them. Commensals compete with harmful bacteria, restricting their growth. In recent years, many products containing these "friendly" bacteria have been promoted as a way of keeping healthy.

Hairs in the nose may not look pretty, but they help to catch dust particles that may be carrying Microorganisms. The sneeze reflex when dust enters the nose also helps, sending the possible invaders flying away at 25 miles (40km) per hour. Mucus in the airways traps microbes, which are then swept up by tiny hairs (cilia) lining the throat and windpipe so that they can be coughed up. Mucus also plays an important part in protecting the vagina and urethra from infection. These two areas also contain helpful commensal bacteria.

sneezes spread diseases
The sneeze reflex is meant to protect us against infection, but the fine droplets sprayed out can spread microorganisms around a room, where they can be breathed in by someone else.

local infections

Despite all these measures, some organisms still get into the body. For example, someone may accidentally cut themselves while weeding the garden. Bacteria in the soil spill into the wound. They start to multiply. At this point, they are confined to a small area of the body—forming a local infection. Local infections are very common and include pimples, athlete's foot, and gingivitis.

To deal with this, the immune system initiates the inflammatory response (see p.14). This comes into effect immediately, and is designed to stop the infection from spreading any further. If many millions of microorganisms are present, a wall of fibrous tissue is also built up around the infected area to prevent the infection from spreading. Inside the closed-off area there is a buildup of dead body cells, dead white blood cells (lymphocytes), and destroyed microorganisms. This is called pus.

when infections spread

In some infections, the microorganism or the toxin that it produces gets into the bloodstream and spreads around the body to produce a systemic infection. The specific immune system then comes into play (see p.15). This takes several days to develop, but is highly effective against most infections. It also results in partial or complete immunity to that particular infection.

Systemic infections always cause a fever. When immune cells are damaged by the invading microorganism, they release a chemical called interleukin-1. This stimulates the brain to raise the body temperature to a level at which most disease-causing microorganisms cannot survive. Unfortunately, a very high fever can also damage the human body. A person suffering a systemic infection also often develops a skin rash. This is due to local areas of damage from the microorganism or the toxins it produces.

Infection is a battle between the immune system and the invading organism. The immune system attempts to wall off and destroy the aggressor, which attempts to avoid death and carry

initial defense
Part of the inflammatory immune response involves phagocytic cells that engulf and destroy foreign cells. This scanning electron micrograph shows a macrophage engulfing Neisseria gonorrhoeae *bacteria (yellow), the cause of gonorrhea.*

evading host defenses

Parasitic microorganisms have developed various ways of evading the body's immune response or hindering its function. Some hide in cells until the immune response against them has waned. Others prevent the immune cells from carrying out their normal actions, and some mutate rapidly, so that they are no longer recognized by the immune system.

infected cell

viral DNA

host DNA

lying low
The herpes virus inserts its genetic material into an infected cell's DNA and waits until the immune response is weakened before replicating itself.

virus mutates (antigenic shift)

flu virus

new strain of flu virus

mutated antigen

disguise
The flu virus undergoes regular mutations that change its surface proteins (antigens), so that it is not recognized by the immune system.

macrophage

bacterium

toxin

suppression
Streptococci *bacteria release toxins that damage immune cells, such as macrophages, preventing them from acting effectively against infection.*

bacterium

macrophage

capsule

nucleus

defense
The bacteria that cause tuberculosis have protective capsules that prevent digestion by destructive enzymes within macrophages.

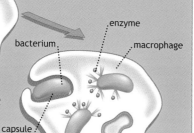

enzyme

macrophage

bacterium

capsule

on living off its host. The outcome depends on whether the parasite can evade or defeat the immune system, and on how virulent (powerful) it is.

evading the immune system

All parasites have evolved offensive or defensive measures against the immune system. Ectoparasites have simply developed methods of piercing the skin. Endoparasites have to be more inventive. Some have evolved to be able to live in the strongly acid environment of the stomach. The bacterium *Helicobacter pylori*, a cause of stomach ulcers, has done this. Others produce toxins that prevent the immune system from functioning effectively, or try to fool or hide from the immune system in one of several ways (see p.18).

Infectious diseases usually become killer diseases in one of two ways. They either successfully beat the immune system, using one of the methods described above, or they are completely new to the body and have such severe effects that they cause irreparable damage before the adaptive immune response has a chance to develop. In the first case, the killer disease may be a chronic one, taking months or even years to disrupt the body to a fatal degree. In the second case, the disease is acute, and may lead to death within hours or days if there is no treatment. In a few rare cases, normally mild infections may kill because the immune system is not functioning properly. This may occur during treatment for cancer or when AIDS has developed in the body.

the white plague

Tuberculosis (TB) is the perfect example of a chronic killer disease. In the 19th century it was called the White Plague and was responsible for about a quarter of all deaths worldwide. Even today, one third of the world's population is infected with the bacterium that causes TB, *Mycobacterium tuberculosis*, and three million people die of it every year.

deadly effects
The effects of the bacterium Mycobacterium tuberculosis *(background image) include progressive destruction of lung tissue, causing chronic coughing of blood.*

Infection occurs when droplets containing the bacteria are breathed in, allowing them to enter the lungs. In 95 percent of cases, the immune system manages to halt the infection. However, the bacteria remain dormant for years, waiting for the immune system to be weakened by age, malnutrition, or a condition such as AIDS. Then they start to multiply, causing progressive lung damage. Larger and larger cavities form in tissue within the lungs, which may eventually result in death. In five percent of cases the original infection is not stopped in its tracks. The bacteria are ingested by macrophages, but prevent themselves from being digested and hijack the immune cells. They then spread throughout the body, using the lymphatic system, and infect many organs, such as the liver and kidneys. Lesions form in all of the affected areas, increasing in size and number. Eventually, one or more of the organs cannot function normally and death occurs if there is no treatment.

the comma-shaped killer

Cholera is a good example of an acute killer disease. It is caused by a comma-shaped bacterium called *Vibrio cholerae* that infects the intestine when water polluted with infected feces is drunk or used in food preparation. The bacteria multiply rapidly, producing a highly toxic poison. The poison causes fluid from the blood to pass into the intestines. Within one to five days, extreme diarrhea develops, sometimes with vomiting. More than 1 pint (0.5 l) of fluid can be lost every hour. In some cases, this can cause death the same day. There is no time for an effective specific immune response to develop if the

swift killer
The Vibrio cholerae *bacterium, shown here in an electron micrograph, is a comma-shaped bacterium with a single flagellum used for movement. The toxin it produces can cause death through dehydration within hours.*

infecting strain of cholera has never been encountered before. Bubonic plague, Ebola, and yellow fever are other examples of acute Diseases that can kill before the specific immune system has time to respond. Fortunately, the survivors of acute infections have partial or complete immunity against another infection of the same type.

transmission of infection

Infectious organisms, being parasites, always need to find new hosts. Most human infections are acquired from other humans. This human-to-human transmission can occur through direct skin contact. (Shaking hands is a more effective way of spreading a cold than sneezing.) It can also occur by breathing in infected droplets of mucus from coughing or sneezing. Millions of virus particles or bacteria encased in tiny protective bubbles can travel down the length of an aircraft or subway car before being inhaled by an unsuspecting and unfortunate person. Infections transmitted this way include influenza, chickenpox, measles, pneumonia, and TB.

Some microorganisms are transmitted through contact with other body fluids, including blood, saliva, and mucus in the genital tract. Higher numbers of sexual partners and greater intravenous drug use have increased the numbers of infections transmitted in these ways. They include HIV, hepatitis B, C, and D, genital herpes, and syphilis.

protective bubbles
Viruses may be carried through the air in microscopic bubbles of fluid that have been sneezed or coughed out.

the kiss
It may be the stuff of romance, but kissing is also a common method of transmitting the virus that causes mono.

contaminated water

Other common methods of transmission involve contamination of water with infected feces. For example, schistosomiasis, caused by a fluke called a schistosome, occurs as a result of schistosome eggs being passed into the water in feces from an infected person. The eggs hatch into larvae that enter and live in freshwater snails. The larvae develop further until they leave the snail and burrow into the skin of someone swimming in the infested water. They then migrate to the bloodstream, where they become adult worms and start the cycle again. Other infections transmitted through feces-contaminated water include giardiasis, amebiasis, hepatitis A, and cholera.

Transmission can also occur through eating infected food. The infection may arise from the food itself being infected (as with chickens infected with *Salmonella* bacteria); from the food being washed with infected water or handled by someone with unwashed hands; or from contamination by insects or rodent droppings.

❝ A successful search for food on the part of one organism becomes for its host a nasty infection or disease. ❞

William H. McNeill, 1976

digging in
A tick burrows headfirst into human skin while feeding. This is irritating enough on its own, but ticks can also spread illnesses, such as relapsing fever and Lyme disease.

the animal route

A few infections are acquired when an infection in an animal jumps the species barrier. These infections are called zoonoses. Anthrax and rabies are killer human zoonoses. Animals can also act as disease vectors. In this case, they do not suffer from a disease themselves, but carry it with them and are capable of infecting humans. Mosquitoes, lice, and ticks commonly carry viruses or rickettsia (small bacteria) that can cause serious infection. For example, epidemic typhus (also called prison fever), which used to be an extremely common fatal infection, is carried by the human body louse. As a louse feeds on a person, it defecates, dropping the rickettsia that cause typhus on to the skin. Because being bitten by a louse makes the person feel itchy, they scratch, infecting themselves as they rub the feces into the skin.

how epidemics develop

Infections can be endemic (constantly present in a population) or occur in epidemics (infecting large numbers of people in waves before disappearing). An epidemic can start only when the nonimmune human population in an area is at a high enough level. Measles provides a good example of this. It is caused by a member of the virus family, paramyxovirus, other members of which are responsible for mumps and influenza. It is highly infectious, and epidemics occur at regular intervals in populations where vaccination levels are low.

Measles is transmitted only between humans, and one case confers lifelong immunity. When an epidemic starts, the disease spreads quickly through all nonimmune people (usually young children). Soon, the numbers of susceptible people are very low. In a small community, infectious people do not have enough contact with susceptible people to continue an unbroken chain of infection. The disease then disappears. Measles will only occur again several years later if someone from outside the area reintroduces the virus. In the new epidemic, only children born since the last outbreak will be infected. In a large community (around 250,000), the pool of susceptible people is large enough for chains of infection to remain. The disease then becomes endemic.

measles
Because measles vaccinations are not given to all children, regular measles epidemics still occur, with a fatality rate of up to 30 percent in developing countries.

key points

• The immune system has three lines of defense.
• Infections can be transmitted by direct contact, infected food and water, coughing and sneezing, or a vector.

killers of the past

Before 10,000 BC, humans lived in very small groups of perhaps two or three families. They lived nomadic lifestyles, moving around in search of food. Despite the harshness of their existence, they did not suffer seriously from disease. They were protected by their low numbers and by the absence of many of the methods of transmission that are common today, such as fecal contamination of water supplies. After the agricultural revolution, all this changed. People started to live in large towns and cities, offering a feast to infectious organisms. The huge changes in human lifestyle heralded an era of new epidemics so dreadful that humans are said to have come close to extinction. Since the 19th century, there have been amazing medical advances as the causes of diseases and their means of transmission were discovered. Vaccines, antibiotics, and antitoxins led to the hope of a disease-free future. Sadly, this has not been realized. New epidemics have arisen, caused, like the ancient plagues, by changes in our lifestyle and in our environment.

black death

There have been many plagues throughout history, the most famous being the Black Death. With thousands dying daily, and no idea of the cause of the horrific disease, the few that could be found to move the corpses tried to protect themselves by smoking, wearing masks, and smearing themselves with foul-smelling potions.

urban ills

Forty thousand years ago, humans had better
health and a longer lifespan than their more
"civilized" descendants were to enjoy for many thousands
of years. The illnesses they suffered from were mainly
chronic (slow-developing), minor infections that still
affect apes and other primates today, such as infections
of the digestive tract. There is also some evidence that
viral diseases such as herpes were common.

Otherwise, people were surprisingly healthy
and well-nourished. Skeletons show that early
humans were tall and had better teeth than
many children in inner city areas today. Men
lived to about 35 years and women to 30.
This may seem short, but in medieval times
the average lifespan for peasants was only
28 years! In any case, most people's lives
were not cut short by disease. The major
causes of death were the hard nomadic life,
accidents, and warfare between rival tribes.

no flies on me
*We inherited more
than our DNA from
our primate
ancestors. Many of
their parasites were
passed on to infest
early humans. We
retain many of
them to this day.*

effects of lifestyle

What protected Stone Age man from disease? Their small
numbers and their lifestyle. A few families lived together
in a tribe, roaming large areas of land in search of new
food sources. The populations were too small for epidemics
to take hold. Constantly moving camp meant that people
did not live surrounded by heaps of garbage and wastes
that could attract pests and parasites. They did not keep
animals that could transmit zoonoses. The diet was
protein-rich, and so their immune systems functioned well.

This all changed with the agricultural revolution, which
began about 12,000 years ago. With the cultivation of

crops came a dependable year-round food supply, and large settlements and cities sprang up in the most fertile regions. Unfortunately, the effects on health were devastating. Skeletons from this period tell a terrible story of rickets, childhood anemia, and other forms of malnutrition—the result of dependence on starchy staple crops. Prostitution, which became a regular trade several thousand years ago, allowed sexually transmitted diseases to flourish. With the general decline in health, immune systems became weaker.

a parasite's paradise

Urban life was (and still is) ideal for parasites. Piles of waste food and feces drew scavengers. Rats invaded

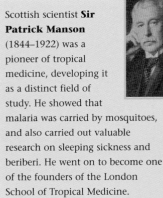

granaries. Besides damaging stored grain, they carried the causes of diseases, including typhus, hemorrhagic fevers, and bubonic plague. Populations were large enough for epidemic diseases such as measles to flourish and become endemic. The overcrowded, dank conditions most people lived in provided a haven for tuberculosis. Water, polluted with the feces of thousands, carried the ever-present threat of dysentery, cholera, salmonellosis, viral hepatitis, and many other intestinal diseases. Water stored in pottery vessels and standing in ditches next to irrigated fields provided the ideal breeding grounds for malarial mosquitoes. About 10 percent of mosquito species transmit diseases to humans. One of these diseases is malaria, which has killed more people than any other illness in history.

perfect dump
Large cities produce vast mountains of rotting garbage that are a paradise for rats and other scavengers.

Scottish scientist **Sir Patrick Manson** (1844–1922) was a pioneer of tropical medicine, developing it as a distinct field of study. He showed that malaria was carried by mosquitoes, and also carried out valuable research on sleeping sickness and beriberi. He went on to become one of the founders of the London School of Tropical Medicine.

malaria

Malaria is spread by the bites of *Anopheles* mosquitoes. The plasmodium parasite infects the liver, then the red blood cells. The cyclical symptoms of fever occur when many blood cells are burst by new parasites erupting out of them.

mosquito bites
A mosquito carrying plasmodia bites a human. The parasites enter in the mosquito's saliva.

establishment of infection
In the liver, the parasite reproduces and develops into a new form that can infect red blood cells.

plasmodia infect blood
The parasites are carried in the bloodstream to the liver.

infected red blood cells

recap

A **zoonosis** is any infectious or parasitic disease of animals that can be transmitted to humans. Zoonoses are usually caught from pets, animals used as food sources, or scavengers. They are also transmitted by insect vectors.

the emergence of zoonoses

By about 4,000 years ago, horses, oxen, goats, and sheep had been tamed. Many people kept them in their tiny houses, risking salmonella and worm infections. Parasites also jumped between the species. Many of our modern infections started off as afflictions of animals.

Anthrax is a zoonosis that became a more common threat with the development of urban life. The cause is a bacterium called *Bacillus anthracis*, which produces spores that can remain dormant for up to 50 years in soil and animal products. People may become infected when they handle material from infected animals. Anthrax was therefore a daily risk to herdsmen, butchers, shearers, wool handlers, and tanners. The most common form of

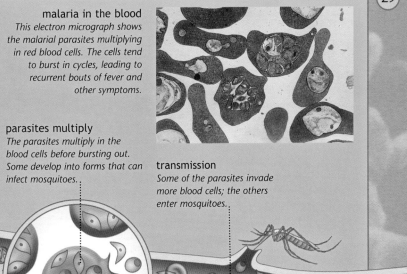

malaria in the blood
This electron micrograph shows the malarial parasites multiplying in red blood cells. The cells tend to burst in cycles, leading to recurrent bouts of fever and other symptoms.

parasites multiply
The parasites multiply in the blood cells before bursting out. Some develop into forms that can infect mosquitoes.

transmission
Some of the parasites invade more blood cells; the others enter mosquitoes.

anthrax (cutaneous) affects the skin, causing black scabs and blisters. Without treatment, the infection may spread through the blood and cause death. A rarer form, pulmonary anthrax, is caught by inhaling spores from animal tissue, and is usually fatal.

Dogs gave humans rabies, worms, and typhus. It is also possible that they gave us measles—the virus is closely related to one that causes distemper in dogs. Horses carried an infection called glanders that causes pneumonia in humans. In fact, many human diseases probably mutated from animal forms, including the smallpox virus and influenza, which began as a sporadic zoonosis. Today, a different version of the flu jumps from an animal reservoir into humans every year.

key points

• Establishment of towns provided ideal conditions for many killer diseases to spread.
• The use of animals in farming allowed zoonoses to be transmitted.

the worst plagues

By 4,000 years ago, people were living in crowded, unhygienic conditions. They were surrounded by sources of infection and were constantly exposed to new microorganisms to which they had no immunity. Everything was ripe for the development of mass epidemics. The emergence of trade and shipping routes and wars fought for territory brought widely separated populations into contact with one another—allowing diseases to spread over a wide geographical area.

Ancient accounts of plagues are common, but are generally unhelpful. Numbers of deaths are unreliable. Doctors had no idea what they were dealing with and the symptoms that they noted are so severe and general that the cause is difficult to to pin down. This is because when a new organism first infects humans, it tends to have very severe effects—often causing high numbers of fatalities. Over many generations, the illness gradually has less severe symptoms and a lower fatality rate.

victim of ancient plague
The Greek physician Hippocrates (460–377 BC), known as the "father of medicine," is depicted here attending a man with the plague.

the athenian plague

The first thorough chronicle of the devastation caused by an epidemic comes from General Thucydides, who kept a detailed record of the plague that attacked Athens in 430 BC. It was the height of the Peloponnesian War. The plague was reported to have started in Ethiopia, and it spread to Egypt and Libya before arriving in Greece by ship. People quickly became feverish and flushed. They sneezed, and their tongues and throats became bloody. Soon they were coughing, vomiting, and suffering from vile diarrhea. Sores that turned into open ulcers broke out all over their bodies. Suffering from unquenchable thirst, the suffering staggered naked through the streets, dying wherever they fell.

General Thucydides (c. 471–400 BC), an Athenian aristocrat, was one of the thousands of people who contracted the plague when it reached Athens in 430 BC. Unlike most, however, he survived the illness and spent most of the Peloponnesian War collecting evidence of the plague for his writings. His famous detailed clinical description of the disease that killed about a quarter of the Athenian land army has led some scientists to suggest that the disease was measles.

The plague seemed to subside and the Athenians launched a fleet against Sparta. Then the disease returned. The Greek hero Pericles died in this bout, and the Athenian fleet was destroyed. The plague continued for five more years, killing at least a third of the city's population. Historians are still uncertain what this plague was, although some have suggested it was an early measles epidemic.

smallpox joins the party

The arrival of smallpox in Europe was probably the cause of the plague of Antoninus. In 165 AD Roman troops sent to Syria to quell a revolt by the locals soon sickened from a new disease unlike any seen before. When the soldiers returned to Rome in 166 AD, they brought the illness with them. The plague spread throughout Europe for 14 years, killing up to seven million people. At one point, 2,000

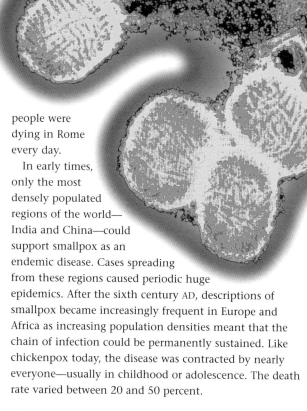

smallpox
This electron micrograph shows smallpox virus particles, which are now thought to survive only in two heavily guarded laboratories. Smallpox started with a flulike illness, followed by a rash spreading mainly on the extremities and face. This developed into pus-filled blisters. Blindness was common in survivors.

people were dying in Rome every day.

In early times, only the most densely populated regions of the world—India and China—could support smallpox as an endemic disease. Cases spreading from these regions caused periodic huge epidemics. After the sixth century AD, descriptions of smallpox became increasingly frequent in Europe and Africa as increasing population densities meant that the chain of infection could be permanently sustained. Like chickenpox today, the disease was contracted by nearly everyone—usually in childhood or adolescence. The death rate varied between 20 and 50 percent.

the defeat of the americas

The more serious effects of a disease on a virgin population were vividly seen in 1520, after smallpox was carried to the Americas. The disease had shown gradual reduction in virulence in Europe. When it reached the New World, it acted as a completely new epidemic. Sufferers were covered in pustules that were so large their flesh came off in chunks whenever they moved. It paralyzed the Aztecs and the Incas, allowing the Spanish to take over with ease. It is estimated that smallpox killed 18 million people in Mexico alone within 100 years.

" A man could not set his foot down unless on the corpse of an Indian. "
Cortés, Spanish conquistador, 1521

the march of bubonic plague

In 542 AD, humans suffered one of the worst plagues in history. It was bubonic plague. This is a horrific and savage killer. The cause is a bacterium called *Yersinia pestis*, which is endemic in rodents and their fleas in many parts of the world. When an infected flea bites a human, the disease is transmitted. After an incubation period of two to five days, fever develops. On the first or second day, buboes (swollen lymph nodes) appear in the armpits, groin, or neck. These can swell to the size of oranges before bursting. The pain of this is enough to drive a patient insane. Before antibiotics were available, more than half of those infected died by the fifth day. Occasionally, septicemia (blood poisoning) caused death within hours. The disease also has a pneumonic form, which usually develops in periods of cold weather. In this form, death is almost inevitable without treatment.

The bacterium that causes bubonic plague, *Yersinia pestis*, is named after the French bacteriologist **Alexandre Yersin** (1863–1943), who discovered it in 1894 and traced the transmission of the disease. Eight years earlier Yersin had collaborated with Emile Roux in research that led to the discovery of the diphtheria toxin.

The outbreak of 542 AD had started in Egypt two years earlier. It spread on trade ships to Constantinople, with terrible effects. Panic, disorder, and murder reigned. Because there were too many corpses to bury, the roofs were removed from the city's fortified towers and bodies were stacked in them like lengths of chopped wood. Soon even the towers were filled and the streets were choked with an overpowering stench. Over 10,000 people died every day. In desperation, the survivors loaded rafts with the dead and set them adrift at sea.

The plague spread rapidly through Europe, recurring frequently until 590 AD. By that time it had killed 50 percent or more of the people in Europe and countless others in the rest of the known world.

the black death

In 1346, another pandemic began. Forty million people were said to have died over the next few years in what people at the time called the Great Dying. Now known as the Black Death, it is thought to have been caused by bubonic plague. (In 2001, a group of researchers claimed that the symptoms were so severe that it was probably not bubonic plague, but a hemorrhagic fever similar to Ebola.)

bubonic plague

The bacteria that cause bubonic plague reproduce rapidly, producing a toxin that destroys blood vessels. As blood leaks from capillaries into surrounding tissues, fluid and blood accumulate, causing swellings. Gangrene develops, causing black lesions on limbs. The bacteria may then enter the lungs or blood, causing certain death.

the rat connection
The bacterium is endemic in wild rodents, spread by flea bites. When too many rats die, fleas may bite urban rats or humans, passing on infection.

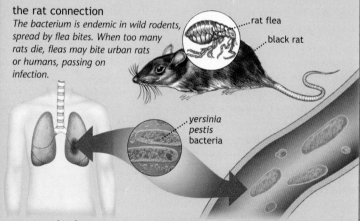

..rat flea

black rat

yersinia pestis bacteria

pneumonic plague
If bacteria enter the lungs, there is severe coughing and bloody sputum. Death is certain without treatment.

bubonic plague
The bacterial toxin results in the formation of swollen buboes, and black marks appear all over the body.

septicemic plague
Sometimes bacteria multipy quickly in the blood. Tiny bruises form all over the body. Death can occur within hours.

flagellants
During the Black Death, people who felt that the plague was a punishment from God marched over Europe, whipping themselves in town squares in an attempt at appeasement. They also carried the plague with them.

As with all previous plagues, this one was carried by traders. In the 1330s, the plague had begun to spread along the caravan routes from Central Asia to China, India, and the Middle East. In 1346, it had traveled as far as the Crimean port of Kaffa (now Feodosiya) on the Black Sea. This city had been under siege for three years by Janibeg, Khan of the Kipchak Tatars. When the Tatars started to die in droves from the plague, Janibeg ordered a withdrawal but is said to have taken a deadly parting shot, using his catapults to propel plague-infected corpses over Kaffa's walls. If true, this tale is the first documented example of germ warfare.

"How will posterity believe that there has been a time when well nigh the whole globe has remained without inhabitants?"

Petrarch, 14th century

Genoese traders who had been trapped in Kaffa rushed to their ships as soon as the city was liberated. They sailed for Italy, carrying the plague with them. By late 1347, it had spread to southern Europe. England was infected in 1348. The death rate of those infected was 70 to 80 percent. In Smolensk, Russia, only five people were left alive.

discovering the cause

No one had the slightest idea of how the disease was spread. It was variously blamed on miasmas, earthquakes, comets, cats, dogs, lepers, gypsies, and Jews. By the 16th

recap

Cholera is caused by the bacterium *Vibrio cholerae*. Carried in feces-contaminated water, it can cause dehydration and death.

infected pump
One of the sources of cholera was found to be an infected public water supply. This contemporary artwork shows Death supplying infected water from a pump.

century, quarantines were enforced, but rats are no obeyers of human laws, and so the plague spread regardless.

In 1894, the French bacteriologist Alexandre Yersin discovered the causative bacterium and the method of transmission was soon deduced. Despite control of rat populations, cases of plague still occur. There are between 10 and 50 cases in the United States annually.

cholera arrives

In 1830, a previously unheard-of disease started to spread west out of India. This was Asiatic cholera, caused by the bacterium *Vibrio cholerae*. Its spread in the 19th century led to worldwide pandemics that killed millions.

The first pandemic stopped short of Europe. Another wave began in 1826 and spread to Russia. In 1830, Moscow had an epidemic that killed 50 percent of those infected. Panicking inhabitants fled the city, spreading the infection widely. It soon reached England. Tens of thousands of people died within days. From England, it spread to Ireland, where immigrants took it to the Americas. Someone visiting Manhattan in 1832 would have seen nothing but carts taking the dead for burial.

At the time, respectable doctors thought that cholera was not contagious. This was because of the results of experiments carried out to discern the cause of yellow fever in Barcelona in 1822. The French doctors who studied the epidemic there decided that the cause was a miasma from decaying organic matter.

However, many lay people, distrustful of doctors, clamored for quarantines. These were worse than useless because no one realized that transmission occurred through water supplies. In London, most drinking water came from the Thames River, which received all of the city's human feces, animal feces, waste from slaughterhouses and hospitals, dead animals, groundwater from cemeteries, and even the odd human

> " In less than six days from the commencement of the [cholera] outbreak, the streets were deserted by more than three-quarters of their inhabitants. "
>
> John Snow, English anesthetist and epidemiologist, 1854

corpse. Everything that could be disposed of down a sewer, was. People drank their own filth, washed in it, and cooked with it. It was no wonder that thousands died every month. Other European cities were no more fortunate.

public reform

By the time of the third cholera pandemic, 10 years later, a sanitary reform movement had started up, led by Edwin Chadwick. He was convinced that disease came from dirt and set out to clean up the unwashed populace of England. He was made commissioner of a new board of health. Laws soon made the government responsible for collecting refuse, building sewers, cleaning the water, and clearing slums. In 1868, when cholera struck again, there were fewer deaths. The success of sanitary reform led to health boards being set up everywhere. Today, a few cases of cholera occur in developed countries every year, but these

Sanitary reformer **Sir Edwin Chadwick** (1800–90) wrote about the appalling living conditions of the working classes in London, and the influence that environment appeared to have on health. Chadwick's three-volume report "Survey into the Sanitary Condition of the Labouring Classes in Great Britain" became a landmark in social history, with its graphic descriptions of how the filth in air, water, and soil was a major factor in the spread of disease in urban areas.

flu pandemic
Spanish flu was more virulent than any we have seen since. It was highly contagious, and all control measures, including the common use of face masks, seemed to be in vain.

are mainly in people who have recently been to Africa or Asia. Eating contaminated shellfish is one of the most common means of transmission.

the forgotten pandemic

In March 1918, a strange form of influenza started to sweep through the United States. It affected people between the ages of 20 and 40, rather than the very young and very old. American troops carried it to Europe and thousands were affected. It circled the globe in about four months. Then a more vicious form appeared and spread around the world again, killing millions. Most of the deaths were from pneumonia, for which there was no drug treatment at the time. All over the United States and other badly affected countries, public gatherings were banned, schools, hospitals, and businesses were closed, and face masks were worn. Nothing helped. In a little over a year, the flu, which was known as Spanish flu, had resulted in one of the worst disasters in history, killing between 15 and 25 million people—more than had died in World War I.

The flu normally has a very low death rate, but so many people are infected that the numbers who die from it are enormous. The virus that causes the flu regularly undergoes changes in its surface proteins—eliminating any immunity to a previous attack. Less often, there is a major shift in the antigens, which usually means that the virus has increased virulence. Many scientists believe that the next great disaster that strikes humans will be a virulent flu pandemic that will kill many millions within months.

key points

• When a particular disease is new to a population, it has more severe effects than normal.
• Public health reform helped eliminate cholera epidemics.

drugs and vaccines

Rubbing onions on warts and then burying them; cutting live pigeons in half and rubbing them on plague buboes; draining blood; applying hot glasses to the skin; drinking concoctions of arsenic; applying egg yolks to wounds: doctors throughout the ages have invented plenty of strange treatments—many useless, some more harmful than the diseases they were meant to treat.

onions for a cure
The ancient cure for warts was to cut an onion in half, rub one half on the wart, and bury it in the ground. As the onion rotted away, the wart was supposed to disintegrate, as well.

old views of disease

These early medical methods may seem laughable to us in this modern age of antibiotics, antiviral drugs, and radiotherapy, but they were the result of a lack of knowledge about what actually caused diseases. There have been many theories about disease over the last 3,000 years. Stone Age man seems to have believed illness to be caused by evil spirits—an idea that still has credence for many tribal peoples today. Other theories held that the movements of stars across the heavens or the appearance of comets preceded epidemics, or that a vile odor given off by rotting tissue could transmit disease. It was not until the development of the microscope in the middle of the 17th century that medical researchers could see for the first time the invisible life teeming all around us. Then the true nature of infection could start to be understood and effective drugs and vaccines produced.

" Garlicke hath powers to save from death, though it makes unsavory breath. "

Regimen sanitatis salernitanum, 1484

modern drugs
Incredible advances in chemistry and biotechnology mean that doctors now have thousands of effective drugs at their disposal in the fight against disease.

our greatest victory

A glimmer of hope for doctors came in 1798, when Edward Jenner (1749–1823) published a book on what he called vaccination. It described his method of protecting people against smallpox. For hundreds of years before, people in Asia and India had practiced inoculation. This involved scratching a small amount of matter from a smallpox sore into the skin of an uninfected person. The usual result was a mild infection that left the patient immune. Sometimes, full-blown smallpox developed, but the benefits of inoculation outweighed the disadvantages.

Jenner's new vaccine developed from his observation that milkmaids rarely suffered from smallpox. He deduced that infection with cowpox, a related virus, provided immunity. No one knew how or why it worked, but it did. Within 10 years, vaccination was used throughout much of the world.

In 1966, the World Health Organization voted for a 10-year mass vaccination campaign to rid the world of smallpox completely. This mammoth task was achieved. The last reported natural infection was in Somalia in 1977. Today, the smallpox virus is thought to exist only in guarded laboratories in Atlanta and Moscow.

cow vaccine
The development of the cowpox vaccine against smallpox was one of medicine's greatest achievements, but it still took nearly 200 years to eliminate the threat of one of our most horrific killer diseases.

cell theory

Despite Jenner's work, the first inklings that disease resulted from abnormalities in the cells and tissues of the body came with the work, published in 1939, of the German physiologists Theodore Schwann (1810–82) and Jacob Schleiden (1804–86). Their microscopic studies showed that all living matter was composed of subunits called cells. This cell theory was elaborated on by another German, pathologist Rudolf Virchow (1821–1902), who stated that all cells originated from other cells and that all disease was a disease of cells.

the work of pasteur and koch

The idea that microscopic organisms were responsible for many common diseases still took a long time to gain a following. Its final acceptance was due in large part to the work of the French bacteriologist Louis Pasteur (whose name is immortalized in the treatment of milk—pasteurization). After early successful work in the structure of crystals, Pasteur was asked to investigate fermentation by a local man who could not understand why some of his beet juice would not ferment properly. At that time, the production of alcohol from sugar was believed to be a purely chemical process.

Through his studies, Pasteur proved that fermentation was really a biological process, caused by the action of yeast. This surprising discovery led him to tackle the age-old problem of spontaneous generation. At this time, in the mid-19th century, it was generally believed that certain forms of life could spontaneously arise. So an apple became moldy because the mold spontaneously appeared on the apple's skin, not because a fungus attacked the apple. Pasteur carried out a series of clever experiments, including carrying sealed flasks containing boiled sugar water to the top of Mount Poupet in the Jura Mountains. There, he opened the flasks to show that in thin air there was little contamination by microorganisms. He went on to show that

recap

Microorganisms include viruses, bacteria, some fungi, and protozoa. They are too small to be visible to the naked eye.

French chemist **Louis Pasteur** (1822–95) discovered that most infectious diseases are caused by germs. Pasteur's "germ theory of disease" is one of the most important in medical history. Pasteur's work became the foundation for the science of microbiology, and one of the cornerstones of modern medicine.

milk mountain
Pasteur's innovative experiments, such as carrying flasks up mountains, were designed to attract attention to his theories about microorganisms.

wine and beer was spoiled by bacterial contamination, and developed a heating method—pasteurization—to prevent this from happening in all sorts of foodstuffs.

early vaccines

In 1880, after isolating the causative organism of chicken cholera, Pasteur developed his first vaccine. He exposed the microbe to the air, causing it to become weakened. Injecting the weakened (attenuated) microbe into chickens provided them with immunity. He then went on to develop a vaccine against anthrax. In 1876, the German Robert Koch had demonstrated that anthrax was caused by a bacterium that could lie dormant for many years. Pasteur carried out a highly public demonstration of his anthrax vaccine, which he had created by attenuating the bacterium. With his reputation at an astronomical high, he also went on to develop a rabies vaccine.

German doctor **Robert Koch** (1843–1910) astounded his parents at the age of five by telling them that he had, with the aid of the newspapers, taught himself to read. In 1870, he volunteered for service in the Franco-Prussian war, and from 1872 to 1880 he was District Medical Officer for Wollstein. It was here that he carried out the research that placed him in the front rank of scientific workers.

Robert Koch also worked to develop a vaccine for tuberculosis. This was less successful than Pasteur's efforts, and actually caused fatalities. However, the vaccine, called tuberculin, became a useful diagnostic tool, because it caused a visible skin reaction in people who had already been exposed to the tuberculosis bacterium.

Today, vaccines are much safer, although they still depend on using attenuated or killed microorganisms, or a modified form of a toxin or antigen, to stimulate immunity. Owing to active vaccination programs, once-common killers, such as measles, mumps, yellow fever, cholera, and hepatitis B, are now much rarer, at least throughout developed countries.

immunization

Immunization protects against particular infectious diseases by using either a vaccine to stimulate the immune response or an immunoglobulin to boost it. This provides long-term protection by stimulating the immune system. In the short term it offers protection against infection, or treatment if infection has already occurred.

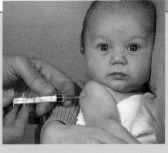

vital protection
A doctor injects a 10-week-old baby girl with the combined diphtheria, tetanus, and whooping cough vaccine (DPT). Injections are the best way of introducing vaccines into the body in order to stimulate active immunization.

active immunization

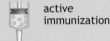

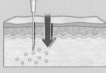

vaccine injected
A person is injected with a vaccine that contains killed or modified forms, or a part of, a disease microorganism.

immune system stimulated
The vaccine stimulates the immune system to produce antibodies against the infection.

immune system ready to respond
If infection with this organism then occurs, there are antibodies present to fight it.

passive immunization

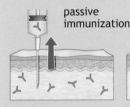
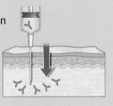

antibodies collected
Blood containing antibodies is taken from people or animals who have recently recovered from the infection.

transfusion
The blood is treated and purified and the resulting serum injected into the person needing protection from the infection.

protection
The antibodies attack the organism if it is present, or provide short-term protection against infection.

the first antitoxins

A major step in the treatment of infectious diseases came in 1890. Two of Koch's former assistants, Emil von Behring and Shibasabura Kitasato, developed treatments for diphtheria and tetanus, using what are known as serum antitoxins. Diphtheria causes a sore throat and fever, and often has fatal complications. Until the 1930s it was a major cause of childhood deaths. Its fatal effects are caused

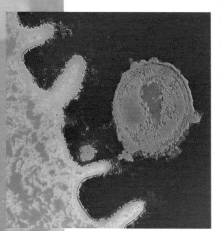

by a toxin released by the bacterium *Corynebacterium diphtheriae*. Tetanus is a serious disease of the central nervous system caused by infection of a wound with spores from the bacterium *Clostridium tetani*. After infection of tissues that are poorly supplied with oxygen, the bacterium produces a toxin that acts on the nerves controlling muscle activity. This results in stiffness of the jaw, and abdominal and back muscles, and profuse sweating. Muscle spasms develop. If these affect the larynx or chest, the patient may suffocate.

diphtheria
This scanning electron micrograph shows a Corynebacterium diphtheriae *(red) in a human throat. When the bacteria multiply, a membrane is formed over the tonsils down to the trachea. This may cause breathing difficulties.*

The antitoxins developed by von Behring and Kitasato involved isolating toxins produced by the bacteria and injecting them into an animal. This animal produced antibodies against the poison. Blood from the animal was then collected and the blood serum prepared for injection into someone suffering from the infection.

Although the mechanism by which the antitoxin worked was not understood at the time, its efficacy was undoubted. The development of diphtheria antitoxin also led to one of the most wonderful stories in medical history—the record-breaking dash by Balto and other sled dogs in 1925 to save children suffering from diptheria in Nome, Alaska.

the age of discovery

The work of Pasteur, Koch, and others after the germ theory of disease was proven led to an explosion in medical discoveries. With understanding of how infections were transmitted, antiseptic measures prevented many common infections after injury, childbirth, or surgery. Public health measures limited infections from food and water supplies. Vaccines could protect against common diseases, antitoxins could cure others. However, the best was yet to come.

In 1909, Paul Ehrlich (1854–1915) developed Salvarsan, the antisyphilitic drug. Syphilis is a gruesome sexually transmitted disease. Initially, sores develop on the genitals, followed by a skin rash a few weeks later. There is then a latent period that can last for up to 25 years. Finally, tertiary syphilis develops, causing tissue destruction, heart valve disease, and progressive brain damage. Paul Erhlich discovered the drug treatment by experimenting with synthetic dyes. He tested some that contained arsenic and found that one killed the organism responsible for syphilis.

A more effective treatment came with Alexander Fleming's discovery of penicillin, which led to the development of many antibiotics. One of the most famous stories in medical history, the discovery occurred by sheer chance. Fleming worked in the Inoculation Department of London's

wonderdog
This statue of Balto stands in New York's Central Park to commemorate the sled dogs who traveled over 600 miles (965km) in six days through the snows of Alaska, carrying vital diphtheria

Bacteriologist **Alexander Fleming** (1881–1955) was born to a Scottish sheep-farming family. He studied medicine at St. Mary's Hospital in London, and published his first report on penicillin's antibacterial properties in 1929. A short man, usually clad in a bow tie, he never mastered the conventions of polite society, but nonetheless was knighted by King George VI in 1944, and shared the Nobel Prize for Medicine with Florey and Chain in 1945.

St. Mary's Hospital. In 1928, a culture of bacteria he was working on became contaminated by mold when spores drifted up the stairs from a laboratory below his while he was on vacation. When he returned, he saw that the bacteria close to the mold had died. He realized that the fungus had produced an "antibiotic" which he called penicillin. In 1935, Australian pathologist Howard Florey found Fleming's report, and he and his colleague Ernst Chain started researching penicillin's effects. By 1943, penicillin was in mass production. Further research led to

drug resistance

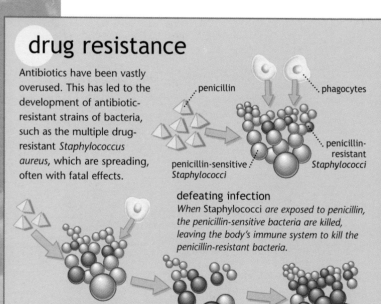

Antibiotics have been vastly overused. This has led to the development of antibiotic-resistant strains of bacteria, such as the multiple drug-resistant *Staphylococcus aureus*, which are spreading, often with fatal effects.

penicillin

phagocytes

penicillin-sensitive *Staphylococci*

penicillin-resistant *Staphylococci*

defeating infection
When Staphylococci *are exposed to penicillin, the penicillin-sensitive bacteria are killed, leaving the body's immune system to kill the penicillin-resistant bacteria.*

low-impact attack
Where the body's immune system is impaired or the quantity of penicillin given inadequate, some penicillin-resistant bacteria survive.

bacteria survive
Within only a few days, the surviving penicillin-resistant bacteria start to multiply in the body.

new colony
The penicillin-resistant bacteria establish a new infection that cannot be treated with penicillin.

the discovery of new antibiotics. Today, they are all produced synthetically.

Antibiotics work in one of two ways. Penicillin drugs and cephalosporins are bactericidal. They kill bacteria by causing the cell wall to disintegrate. Water then enters the bacteria unhindered, making it swell and burst. Tetracyclines and aminoglycosides are bacteriostatic. These do not kill the bacteria. Instead, they inhibit bacterial growth, giving the immune system time to mount a defense.

key points

• Cowpox provided immunity to smallpox.
• Antibiotics kill bacteria or inhibit their growth.
• Antitoxins provide passive immunity.

transfer of resistance

Bacteria have a ring of DNA called a plasmid that carries genetic coding for the production of drug-resistant substances. Bacteria can pass these genes on to other bacteria in a process called conjugation. A small tube called a pilus forms from the donating bacteria and joins to the recipient.

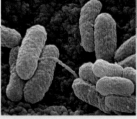

conjugation

One of these Salmonella *bacteria in the intestine has formed a pilus for transfer of DNA to another.*

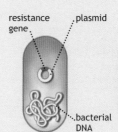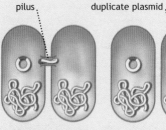

resistance gene · plasmid · pilus · duplicate plasmid ·

····bacterial DNA

plasmid DNA

The plasmid is a ring-shaped piece of DNA, separate from the bacterial DNA, with different genes.

plasmid transfer

A small tube called a sex or conjugation pilus sprouts out from one bacterium, joining it to another.

new resistance

The plasmid duplicates and passes through the pilus, conferring resistance on the recipient bacterium.

modern killers

New diseases usually emerge when we change our lifestyle or when we start to occupy new territories. The plagues of the past were related to the colonization of new countries, the establishment of new trade routes, and the development of agriculture, animal husbandry, and urban life. Over the last 100 years, our lifestyle (especially in the developed world) has altered dramatically, and we have caused more rapid environmental change than at any other period in Earth's history. These factors have led to the emergence of new and even more deadly diseases. Heart disease, cancer, and AIDS have become the plagues of modern times. Older diseases that once seemed defeated, such as tuberculosis, have returned in deadlier, drug-resistant forms. Developments in genetic engineering have raised the specter of man-made epidemics. Based on these factors, it is possible to predict how some disease trends will change in the next few years, so that we can anticipate and prepare for potential killers of the future.

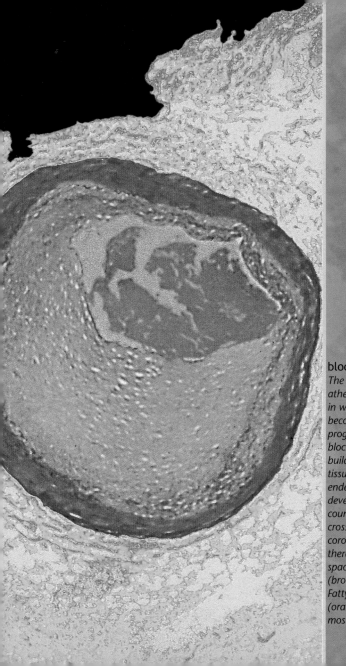

blocked artery
The condition atherosclerosis, in which arteries become progressively blocked by a buildup of fatty tissue, is now endemic in developed countries. In this cross-section of a coronary artery, there is very little space for the blood (brown) to flow. Fatty tissue (orange) blocks most of the artery.

modern threats

New diseases are constantly appearing. Some of them are mild; others have catastrophic effects. The emergence of a new disease can occur for many reasons: a mutation or adaptation in an existing microorganism, the transmission of an animal infection to humans, or even just a change in our lifestyle.

the emergence of polio

Poliomyelitis is a viral disease that, in severe cases, attacks the brain and spinal cord, leading to extensive paralysis. Although the first documented case of polio occurred in the late 18th century, cases for the next 100 years were few and far between, and the first epidemic started as recently as 1907. In 1916, there was a large outbreak in the United States. Within a year 6,000 people had died. Epidemics continued to occur, with worse effects and affecting increasingly older people, until the mid-1950s, when the Salk vaccine was developed.

Changes in our lifestyle meant that polio suddenly became a killer. The polio virus spreads through contamination of water or food with feces. The virus

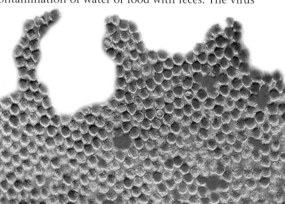

polio virus
Thousands of polio virus particles cluster together in this electron micrograph. They are the cause of one of the last century's most dreaded diseases—poliomyelitis.

multiplies in the intestine, usually causing only a minor infection, after which the infected person is immune for life. In a few cases, the virus travels through the blood to the brain and spinal cord, causing paralysis. The older a person is when they contract the infection, the greater the likelihood of this occurring. In the crowded, dirty conditions of cities in developing countries, infection was common by the age of three. Serious disease was therefore rare. In developed countries, with the development of sanitation and easing of overcrowding, children often did not become infected until much later. They were then more likely to suffer severe effects. Polio's deadly effects were, ironically, due to our healthier lifestyle.

disease named after victims

In 1976, another killer disease suddenly appeared. The American Legion Department of Pennsylvania was holding its annual convention at a hotel in Philadelphia. Within two weeks, 149 of the attendees came down with the same puzzling illness, characterized by fever, coughing, and pneumonia. By the fall, 221 people had become sick and 34 had died.

It took six months for investigators to discover the cause: a previously unknown bacterium, which was named *Legionella pneumophila*. The disease was officially called legionellosis. A year later, new epidemics started in California and Great Britain, killing a quarter of those infected. More cases followed. Strangely, they were all linked to offices, hotels, and hospitals. The link between all these buildings was their air-conditioning systems. It became clear that the bacterium multiplies in air

modern slums
Standards of hygiene and overcrowding in shanty towns in developing countries are worse than in the earliest cities. Conditions for the emergence of epidemics and new illnesses are ideal there.

coronary artery disease

Atherosclerosis is a buildup of fatty plaques called atheroma on the walls of arteries. When atherosclerosis occurs in the coronary arteries that supply the heart, the heart becomes short of oxygen and cannot work efficiently. If one artery becomes completely blocked, part of the heart muscle will die, causing a heart attack to occur.

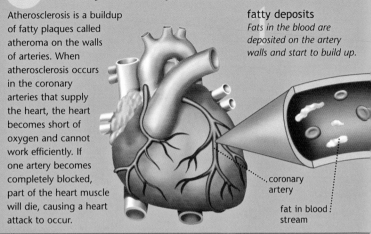

fatty deposits
Fats in the blood are deposited on the artery walls and start to build up.

coronary artery

fat in blood stream

conditioners and is spread in aerosol droplets. *Legionella pneumophila* has also been found in cooling towers and hot-water systems. As the number of air-conditioning systems in buildings around the world increases, more epidemics of this disease will probably occur.

heart disease epidemic

Changes in our lifestyle and our longer lives have led to an enormous increase in what was once a rare disease: atherosclerosis. This is the buildup of fatty plaques on the inside walls of arteries throughout the body, which can lead to heart attacks and strokes. We often get little or no exercise, we eat food high in fat and cholesterol, we smoke, and we suffer from more stress than at any time in history. These factors all make atherosclerosis more likely.

Atherosclerosis is a silent killer. There are no symptoms until blood flow through one or more arteries is so disrupted that there are severe effects. The effects are most catastrophic when one or more coronary arteries, which

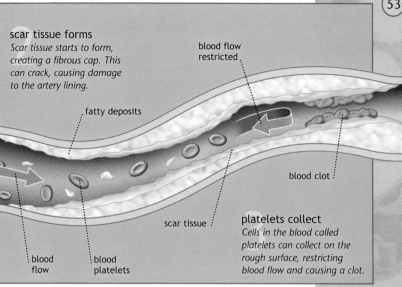

scar tissue forms
Scar tissue starts to form, creating a fibrous cap. This can crack, causing damage to the artery lining.

blood flow restricted

fatty deposits

blood clot

scar tissue

blood flow

blood platelets

platelets collect
Cells in the blood called platelets can collect on the rough surface, restricting blood flow and causing a clot.

supply the heart muscle with blood, become narrowed or blocked. The first symptoms of coronary artery disease (atherosclerosis of these arteries) are either angina or a myocardial infarction (heart attack), depending on the extent of the blockage. Coronary artery disease is now the biggest killer in the Western world.

Atherosclerosis also makes the development of blood clots in blood vessels more likely. These blood clots (emboli) can either block the blood vessel they form in, or move through the blood to block a blood vessel elsewhere. This can lead to strokes and pulmonary embolisms (blood clots blocking arteries in the lungs). Strokes can also occur when arteries supplying the brain are narrowed and blocked by atherosclerosis. There has been increasing concern recently about a condition called economy-class syndrome. Travelers sitting in cramped conditions on airliners may develop a blood clot as blood pools in their legs. Many people have suffered from a pulmonary embolism after taking a long-haul flight.

how cancers develop and spread

Development of cancer is due to damage to a cell's genetic material. Genes that control cell division (called oncogenes) can be damaged by carcinogens, viral infection, or simply by inheritance of a faulty gene. The cell then becomes immortal and divides uncontrollably.

cancer cells form
If damage to the oncogenes goes on over a long period of time, such as when someone is a long-term smoker, the repair mechanisms may not be able to correct the damage. The cells then becomes cancerous and multiply rapidly.

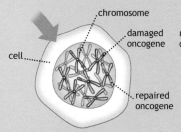

chromosome

damaged oncogene

cell

repaired oncogene

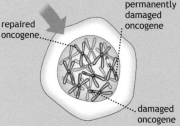

permanently damaged oncogene

repaired oncogene

damaged oncogene

damage from carcinogens
Carcinogens affect the cell, damaging genes on the chromosomes. Newly damaged oncogenes are usually repaired rapidly by the cell.

damage accumulates
The effects of repeated damage build up. Over time, repair mechanisms also become less efficient or may be damaged themselves.

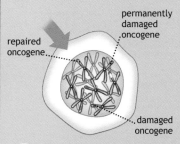

permanently damaged oncogene

repaired oncogene

damaged oncogene

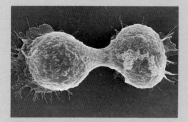

cell becomes cancerous
If enough oncogenes are not repaired, the cell can become cancerous. Its structure changes from that of the normal organ's cells and it divides more rapidly.

dividing cancer cell
This scanning electron micrograph shows a breast cancer cell dividing. The two daughter cells are still connected by a narrow bridge. Breast cancer is the most common cancer in women.

tumor develops
Cancer cells have large nuclei and do not resemble the cells that they originated from. The tumor they form may erode epithelium (cell tissue), causing ulcers. Bleeding may occur when cancer cells break through blood vessel walls.

tumor spreads into surrounding healthy tissue as cancer spreads

healthy tissue

cells in the tissue invaded by faster-growing cancer cells

irregularly shaped cancerous cells divide uncontrollably

tumor breaks through blood vessel, allowing cancerous cells to spread through bloodstream

cancerous cell carried by blood to another area of body

bleeding, caused by cancerous cells breaking down the walls of blood vessels in tumor

cancerous cells break through blood vessel at new site and lodge there

cancer spreads
When cancer cells breach a blood vessel or lymph vessel, they are carried through the body. They then break out of the blood vessel at another part of the body and form a secondary tumor there.

cancerous cells start to divide in new site

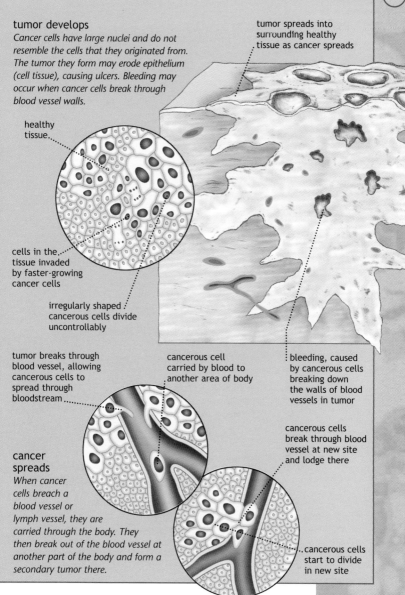

deadly cancers

Cancer gets its name from the Greek word for crab, because the Greek physician Hippocrates (460–377 BC) thought that a spreading malignant tumor resembled a

crab's claw. Any organ in the body can become cancerous, and cancers have become more common due to changes in our lifestyle (sunbathing, diet, and sexual activity have been linked to some cancers), our increasing lifespan (cancers are more likely to develop with increasing age), and exposure to more sources of radiation and carcinogens.

smoking and cancer
Tobacco smoke contains many harmful chemicals, some of which are carcinogenic. Most lung cancers are due to the effects of these carcinogens.

Although cells are constantly exposed to cancer-causing factors, they rarely become cancerous because cells have mechanisms for repairing damaged genes, and more than one gene has to be damaged for cancer to develop. This is why cancers are more common in older people—the cells have more time to accumulate damage. The immune system, which normally destroys cancer cells, also weakens with age.

the AIDS epidemic

HIV-infected T cell
This scanning electron micrograph shows a T cell (green) infected with HIV. As the virus replicates, new virus particles bud out from the T cell, weakening and eventually killing it.

The worst new killer disease of the 20th century was, without a doubt, AIDS. This disease, caused by human immunodeficiency virus (HIV), is believed to have originated in chimpanzees, which suffer from a similar virus called SIV (simian immunodeficiency virus). Perhaps SIV mutated into a form capable of infecting humans. Transmission between ape and human may have occurred because of hunting and skinning apes by humans, or through blood-sucking stable flies. The epidemic spread widely because of changes in human lifestyle.

In the early 1980s, doctors in New York and California noticed that some gay men were suffering from very rare disorders, including Kaposi's sarcoma (a skin cancer) and a rare pneumonia called PCP. The patients' immune systems seemed overwhelmed, leaving them open to infections. The number of cases rose rapidly. In Europe and the US, the disease mainly affected homosexual men, intravenous drug users, and hemophiliacs. In Africa, it seemed to be spread by heterosexual intercourse.

AIDS is caused by HIV, which is a retrovirus (its genetic material is RNA instead of DNA). Over 40 million people are thought to have been infected to date, and the numbers are still rising sharply. Fortunately, the virus is not airborne. It can only be spread by direct contact with blood, semen, or vaginal fluids, but it can also infect embryos across the placenta.

HIV's spread has been accelerated by the use of hypodermic needles, blood transfusions, organ donations, the movement of migrant workers, and changes in sexual behavior. Potential vaccines are being tested, but the projected death rates are staggering. At least 30 million people will die in Africa alone over the next decade. In some African countries, a quarter of the population is infected. Treatment and prevention are difficult because the virus mutates regularly in the body after infection. The main worry for epidemiologists is that the HIV virus may mutate to an airborne form. However, terrifying though that prospect is, it may not be the greatest threat to our future.

key points

• New diseases emerge or become more severe with changes in lifestyle or environment.
• Coronary artery disease has increased due to lifestyle changes.

deadly addiction
The growing incidence of intravenous drug abuse has allowed the rapid spread of blood-borne infections, such as hepatitis and HIV.

how HIV works

The HIV virus is a retrovirus that contains a chemical called reverse transcriptase. The virus primarily infects T lymphocytes (see p.15). Once inside a T cell, the virus uses reverse transcriptase to convert its RNA into DNA. This viral DNA then hides in the DNA of the host cell where it may remain dormant for years. Then it starts to replicate, killing the T cell in the process. The immune system gradually becomes compromised, leaving the patient vulnerable to a wide range of infections.

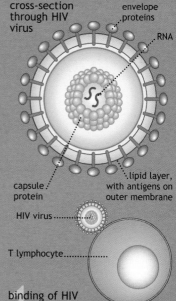

cross-section through HIV virus

envelope proteins

RNA

capsule protein

lipid layer, with antigens on outer membrane

HIV virus

T lymphocyte

infection of immune cells
When HIV enters the blood, it infects blood cells that have a protein receptor called CD4 on the surface. These cells are mostly CD4 T lymphocytes, which are responsible for fighting infections.

HIV virus fuses with T cell

nucleus

T-cell membrane

RNA

HIV virus crosses into T-cell nucleus

1 binding of HIV to T cell
The HIV virus recognizes and binds specifically to the protein CD4 receptor, which is present in this case on a T lymphocyte.

2 penetration at cell surface or uptake into vacuole
The viral membrane fuses with the T cell's membrane, allowing the viral RNA to be released into the cell. Sometimes, the whole virus is absorbed into the cell before the RNA is released into the nucleus (lower image).

symptoms of HIV

The initial infection with HIV may be symptomless, or there may be a short-lived flulike illness with fever and perhaps a skin rash. Many people then have no symptoms until their immune system becomes extremely weakened and they develop a severe illness. Other people might experience recurrent mild infections, such as *Herpes simplex* infections, colds, chest infections, weight loss, lethargy, and dry, itchy skin, before becoming seriously ill.

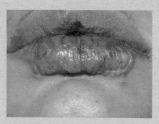

cold sore
Recurrent, severe cold sores caused by the Herpes simplex *virus may be one of the earliest symptoms of HIV infection.*

invasion of host DNA
Within 12 hours, viral RNA is converted into viral DNA by the enzyme reverse transcriptase. The viral DNA then incorporates into the cell's own DNA. After a variable time, the virus initiates production of new RNA.

nucleus
viral RNA

the action of viral RNA
The viral RNA is carried into the cell's cytoplasm where it is able to produce viral proteins, which will form the basis of new HIV viruses.

viral proteins

assembly complete
Many viral components are produced and the viral particles are assembled to form many copies of the original infecting virus.

new HIV viruses

cell destruction
The new viruses bud out of the T cell, weakening it in the process. The viruses then go on to infect other T cells and repeat the cycle. The T cell dies soon afterward.

new virus
dying T cell

potential pandemics

One of the most worrying groups of diseases for researchers today are the hemorrhagic fevers. These illnesses, although rare, are among the deadliest human infections. They include Ebola, dengue hemorrhagic fever, Junin virus, Venezuelan hemorrhagic fever, Rift Valley fever, and Lassa fever. The effects of these diseases include a fever, headache, and muscle pains, followed by the leaking of blood through the blood vessel walls. Blood pours out of internal organs and the nose, mouth, anus, and eyes. This leads to shock, which is often fatal. Some of these viral diseases are carried by insect vectors, others by infected animals.

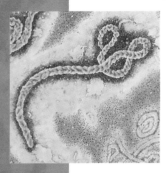

ebola
The Ebola virus (shown in this artwork) is one of the world's most feared infectious agents. It causes hemorrhagic fever, with a very high fatality rate. Scientists are concerned that it may soon cause a major epidemic or pandemic.

Rift Valley fever is a disease of cattle and sheep, found in the Rift Valley area of Africa. In 1977, a human epidemic of the disease suddenly started in Egypt. No one knows how or why, which means that it could probably happen again. Hundreds of thousands of people became sick and thousands died.

level 4 viruses

Junin virus was first identified in 1953 in Argentina. It is carried by a fieldmouse, and people become infected by breathing in dust that has been contaminated by mouse droppings. The virus is classified Biosafety Level 4, as is HIV—there is no vaccine and no cure. About 25 percent of those infected die within days.

Ebola, another Biosafety Level 4 virus, is the deadliest human infection besides rabies and HIV. An epidemic in Sudan in 1976 killed more than half of those infected. Fortunately, it subsided. Nothing is known of how this virus is transmitted, which makes control measures very difficult. Researchers are deeply worried that Ebola, or one of the other hemorrhagic fevers, will become easier to transmit. Then it could spread in a pandemic, killing many millions of people.

bioterrorist attack

The United States started research into biological agents for warfare during World War II. Anthrax and botulism were the first candidates investigated. Both of these deadly diseases are caused by hardy bacteria that have short incubation periods. The goal was to develop anthrax and botulin bombs that could be dropped on Germany if it initiated a biological attack. At Porton Down in England, similar research was undertaken. During the war, British scientists released *Bacillus anthracis* bacteria on Gruinard Island off the coast of Scotland, after they had grouped live sheep into areas to assess the virulence of the bacteria. The island remains uninhabited. There was also deep suspicion that the Russians were working on developing their own bioweapons.

Although the US and more than 100 other countries signed the 1971 Convention on Biological and Toxin Weapons, which involved agreeing to destroy stockpiles

recap

Anthrax is a bacterial infection of livestock that can spread to humans. It is caused by *Bacillus anthracis* and can occur as cutaneous or pulmonary anthrax. The latter occurs when anthrax spores are inhaled, and is usually fatal.

anthrax island

In 1941–42, sheep were tethered in marked-out areas on Gruinard island, and anthrax spores were released. The island remained contaminated with anthrax spores for over 50 years, and is still uninhabited.

decontamination
foam testing
*Decontamination
foams have been
developed to
neutralize all
chemical and
biological agents
over a large area.*

of biological weapons, there is increasing suspicion that some countries, including rogue states, have pushed ahead with bioweapon programs.

Besides botulism and anthrax, potential agents include *Yersinia pestis* (the cause of bubonic plague) and smallpox. After the eradication of smallpox, stores were kept in highly guarded locations in the US and Russia. There are fears that some Russian stocks may have been stolen and used to develop weapons of mass destruction. If smallpox were released today, the effects would be unimaginably horrific.

Genetic engineering has also opened up the possibility of producing a virus such as the flu that is vastly more virulent than normal strains, or even of producing a biological weapon that affects only a certain ethnic group. A few years ago, the threat of a pandemic caused by a biological attack seemed unlikely. However, the anthrax attacks in the US in 2001 after the September 11 terrorist attack alerted people to the potential threats. It is highly likely that something similar will happen again.

key points

- Hemorrhagic fevers are caused by viruses.
- Anthrax and botulism have both been investigated as possible agents for use in biological attacks.

a mutant strain

No century in documented history has gone by without new diseases appearing. Why should the 21st century be any different? In many cases, the "new" illness resulted from a mutation in an existing microbe. The Spanish flu pandemic of 1918 resulted from a new strain of flu virus. Flu researchers are adamant that it is not a case of "if" but "when" a new flu strain will have similar effects. In the 1980s, the bacterium *Haemophilus influenzae*, which causes meningitis, mutated to form a new deadly strain.

future disease trends

◄ recap

Tuberculosis is caused by the bacterium *Mycobacterium tuberculosis*. It is a chronic disease that progressively attacks the lungs or other organs. Strains have developed that are resistant to most of the drugs used in treatment.

Besides the emergence of new diseases, many infections that seemed to have been defeated by medical advances have been occurring in increasing numbers in recent years. It is likely that diseases such as tuberculosis, which have developed drug-resistant strains, will become more common. With ineffective drug treatments, deaths from these illnesses will increase enormously.

Other diseases that had become rare because of vaccination, such as measles, are now starting to reemerge as worries are voiced over the safety of vaccines or as vaccination programs break down because of wars. It is likely that new epidemics of measles and mumps will occur in the next few years.

food poisoning

Thousands of people in the US and Europe still die every year from food poisoning caused by *Salmonella* and *Shigella* bacteria. Cases of listeria continue to increase. Despite our raised awareness of food hygiene, the danger is actually higher because of increasing automation of the food industry and a greater reliance on carryout food.

A dangerous new strain of the common *Escherichia coli*, or *E. coli*, bacterium (0157:H7) has already caused many deaths, and more epidemics will doubtless occur. There is also the chance that BSE (commonly called mad cow disease) has spread from cattle to the human population, causing a disease called new variant

sickness to go? *Quick and cheap, but the food source and the hygiene standards may be questionable.*

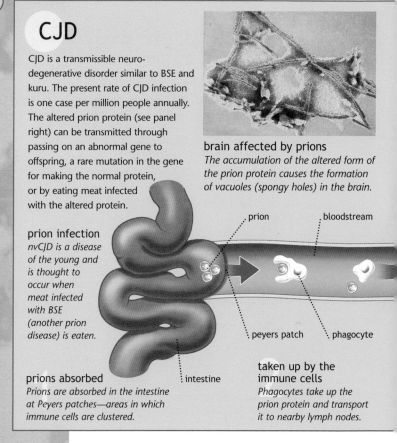

CJD

CJD is a transmissible neuro-degenerative disorder similar to BSE and kuru. The present rate of CJD infection is one case per million people annually. The altered prion protein (see panel right) can be transmitted through passing on an abnormal gene to offspring, a rare mutation in the gene for making the normal protein, or by eating meat infected with the altered protein.

brain affected by prions
The accumulation of the altered form of the prion protein causes the formation of vacuoles (spongy holes) in the brain.

· prion   · bloodstream

prion infection
nvCJD is a disease of the young and is thought to occur when meat infected with BSE (another prion disease) is eaten.

prions absorbed   : intestine
Prions are absorbed in the intestine at Peyers patches—areas in which immune cells are clustered.

· peyers patch   · phagocyte

taken up by the immune cells
Phagocytes take up the prion protein and transport it to nearby lymph nodes.

recap

Malaria is a disease caused by a protozoan parasite that infects red blood cells. Infection is passed from infected mosquitoes.

Creutzfeldt-Jakob disease (nvCJD). The incubation period for this disease could be as long as 20 years, and so no definite estimates of this disease's impact can yet be made.

climate change

It also appears that the world is getting warmer. If climate change does occur, the ranges of vectors such as mosquitoes will greatly increase. Malaria is already spreading north to the Mediterranean region. It is possible

what are prions?

CJD is a prion disease, caused by an altered form of the cellular prion protein (PrPC), which is found on nerve cells. The function of the normal protein is not known. It has been suggested that it plays a role in neural differentiation, or sleep regulation, or acts as a cell–cell adhesion molecule in peripheral tissue.

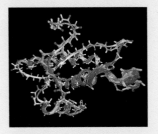

abnormal prion protein structure

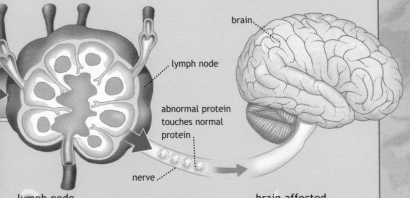

brain

lymph node

abnormal protein touches normal protein

nerve

lymph node
Normal prion proteins on nerves in the node are altered when touched by the abnormal prion. The altered proteins propagate up the nerve to the brain.

brain affected
The abnormal protein accumulates in the brain and holes (vacuoles) form.

that illnesses such as yellow fever may also see an expansion of their range. What seems certain is that we will never be completely free from disease. With every medical advance we make, there will be a new disease or strain to tackle. Changes in the way we live and affect our environment make an increase in our disease burden more rather than less likely. However, we are nothing if not ingenious. For every new threat that comes our way, we will strive to find a cure.

key points

• Drug resistance will become a greater problem in the next few years.
• Fatal food poisoning will become more common.

glossary

anthrax
A deadly bacterial infection of livestock that can spread to humans.

antibody
A protein produced by B lymphocytes to neutralize an antigen in the body. Antibodies are formed after natural infection or the administration of a vaccine.

antigen
Any foreign substance that can trigger an immune response. Antigens are usually proteins not found in the body, such as the surface proteins of bacteria and viruses.

antitoxin
A preparation used in passive immunization, consisting of antibodies from an animal that has been exposed to a particular bacterial toxin.

atherosclerosis
A disease in which the inner wall of arteries is thickened by plaques of fatty tissue. Blood flow through affected arteries is reduced.

bacteria
A group of single-celled organisms, some of which cause human disease. Disease-causing bacteria are classified as: cocci (spherical), bacilli (rod-shaped), and spirochetes (spiral-shaped).

B lymphocyte
A type of white blood cell; part of the immune system. B lymphocytes transform into plasma cells, which secrete antibodies.

cancer
Any of a group of diseases in which cells grow uncontrollably in the body.

cholera
An infectious disease of the intestine, caused by the bacterium *Vibrio cholerae*. Symptoms include profuse diarrhea.

diphtheria
A bacterial disease, caused by *Corynebacterium diphtheriae,* which causes a sore throat and fever, and may be fatal.

ectoparasite
A parasite that lives in or on its host's surface.

endemic
A disease or medical condition that is constantly present in a particular region or group of people.

endogenous disease
Any disease arising within the body.

endoparasite
A parasite that lives inside its host, in its cells or internal organs. Endoparasites include some bacteria and all viruses.

epidemic
A disease or medical condition that is not generally present in an area or population but occasionally affects a large number of people over a relatively short period of time.

exogenous disease
A disease caused by an external factor, such as an infectious agent or exposure to a toxic chemical.

fungus
One of many different species of simple life forms. Fungi include yeasts, molds, mushrooms, and toadstools. They can be single-celled or occur in chains of filaments called hyphae. The nails, genitals, and skin are the most common sites of fungal infection in humans.

HIV
Human Immunodeficiency Virus. HIV invades specific human T cells, weakening the immune system. After a variable length of time, this leads to the development of AIDS.

immune system
A complicated system of cells and proteins that acts to protect the body

against infection and the development of cancer.

immunization
A method of inducing immunity by administering a vaccine. The terms "vaccination" and "immunization" are used interchangeably.

infectious agents
Any agent capable of causing a disease that can be transmitted. Infectious agents include bacteria, viruses, fungi, protozoa, and prion proteins.

inflammatory immune response
Part of the immune response, involving phagocytes and the release of proteins that act to dilate blood vessels and make them more permeable. This causes redness and swelling.

lymph node
A small organ lying in a lymphatic vessel. Lymph nodes house lymphocytes and macrophages.

lysozyme
An enzyme in tears, saliva, sweat, and many tissues that breaks down bacteria.

macrophage
A type of white blood cell; part of the immune system; phagocyte.

mutation
A change in a gene that may cause the gene to behave in new ways.

pandemic
A widespread epidemic that occurs over a large geographical area, affecting a large proportion of the population.

passive immunization
The process of inducing immunity to an infection by administering antibodies from a person or animal who has recovered from that infection.

phagocyte
A cell that can surround, engulf, and destroy microorganisms, dead cells, and foreign particles in the body.

plague
A term used to refer to any serious killer pandemic, or, more correctly, to the disease caused by the bacterium *Yersinia pestis*.

polio
The short name for poliomyelitis; an infectious disease caused by a virus that usually only causes mild illness, but can lead to paralysis.

prion protein
An infectious agent that is a protein. Prions do not contain genetic material and are not living. They cause diseases that involve degeneration of the nervous system and have a long incubation period.

smallpox
A highly infectious viral disease that has been totally eradicated worldwide. The illness involved an influenza-like episode, followed by the development of a rash that formed pus-filled blisters.

T lymphocyte
A type of white blood cell; part of the immune system. There are three main types: killer cells attach to infected or abnormal cells and release chemicals to destroy them; helper cells enhance the activity of killer cells and B lymphocytes; and suppressor cells turn off the immune system.

toxin
A poisonous protein produced by pathogenic bacteria, venomous animals, or poisonous plants.

vaccine
A preparation given to induce immunity against a specific infectious disease. A vaccine usually contains part or all of the disease-causing organism.

vector
An organism that carries an infectious disease from one individual to another. Vectors can carry infections between individuals of the same species, or from one species to another.

virulence
A microorganism's ability to cause disease.

virus
The smallest type of infectious agent, with a simple structure of genetic material surrounded by a protein coat.

zoonosis
A human infection that normally occurs in an animal species.

index

further reading

The Black Death: a History of Plagues, William G. Naphy and Andrew Spicer, Tempus Publishing, 2001.

Yellow Fever, Black Goddess: The Coevolution of People and Plagues, Christopher Wills, Addison Wesley, 1997.

Microbe Hunters, Paul De Kruif, Harcourt, 1996.

Future Plagues, Peter Brooke-smith and Roy Porter, Cassell Illustrated, 1999.

How the Immune System Works, Lauren M. Sompayrcie, Blackwell Science, 2001.

Virus X: Understanding the Real Threat of New Pandemic Plagues, Frank Ryan, HarperCollins, 1996.

Rats, Lice and History, Hans Zinsser, Penguin books, 2000.

The Cambridge Illustrated History of Medicine, Roy Porter (editor), Cambridge University Press, 2001.

internet resources

http://www.beloit.edu/~biology/emgdis/history.emgdis
Links to sites that give both the history of certain plagues and their impact on societies.

http://www.mic.ki.se/HistDis
Links to articles covering the history of major plagues and the evolution of diseases.

http://www.cdc.gov
Provides up-to-date information on all diseases and current epidemic status.

acknowledgments

I would like to thank everyone at DK who worked on this book, especially Peter Frances, Jonathan Metcalf, and Phil Ormerod. Warmest thanks go to Jane Laing for her advice, friendship, and editing skills, and to Christine Lacey for creating a beautiful book with only a few rough sketches to guide her. I would also like to thank John Gribbin for his helpful comments and Eben Arnold for his advice and inspiration.

Jacket design: Nathalie Godwin; **proof-reading**: Jane Simmonds

picture credits

The publisher would like to thank the following for their kind permission to reproduce their photographs. KEY: SPL = Science Photo Library

1: Alfred Pasieka/SPL; **2**: NIBSC/SPL; **4** background: Prof S.H.E. Kaufmann & Dr. J.R. Golecki/SPL; **5**: SPL; **6**: Bob Elsdale/Getty Images; **8**: David Scharf/SPL; **9**: CNRI/SPL; **10** left: NIBSC/SPL; **10** right: Kwangshin Kim/SPL; **11** left: Sinclair Stammers/SPL; **11** right: E. Gueho/CNRI/SPL; **13**: Dr. Kari Lounatmaa/SPL; **17**: Juergen Berger, Max-Planck Institute/SPL; **19** background: Institut Pasteur/CNRI/SPL; **20**: Eye of Science/SPL; **21**: Le Baiser, Auguste RODIN, Inventory number: S.1002, Material: Marble, Musee Rodin, Paris/Dorling Kindersley; **22**: Volker Steger/SPL; **23**: Lowell Georgia/SPL; **24**: CNRI/SPL; **25**: Brian Wilson/Ancient Art & Architecture Collection; **27** top: Weiss, Jerrican/SPL; **27** below: SPL; **29**: Dr Gopal Murti/SPL; **30**: AKG London; **31**: Bettmann/Corbis; **32**: Alfred Pasieka/SPL; **33**: Leonard de Selva/Corbis; **34** below center: CNRI/SPL; **35**: Bibliotheque Royale, Brussels. Photo: AKG London; **36**: SPL; **37**: SPL; **38**: Bettmann/Corbis; **41** center: Custom Medical Stock Photo/SPL; **42**: National Library of Medicine/SPL; **43**: Saturn Stills/SPL; **44**: Dr Kari Lounatmaa/SPL; **45** below: SPL; **47**: David Scharf/SPL; **50**: Institut Pasteur/CNRI/SPL; **51**: Peter Menzel/SPL; **52**: Cecil H Fox/SPL; **53**: SPL; **54**: SPL; **56**: Oscar Burriel/SPL; **56** background: Scott Camazine/SPL; **57**: Alain Dex, Publiphoto Diffusion/SPL; **59**: SPL; **60**: A. Gragera, Latin Stock/SPL; **61**: Martin Bond/SPL; **62**: Randy Montoya/Sandia National Laboratories/SPL; **64**: EM Unit, VLA/SPL; **65**: Alfred Pasieka/SPL; **68**: SPL; **70**: SPL. All other images © Dorling Kindersley. For further information see: **www.dkimages.com**